Irfan Jamil et al.

Application and Composition observing system of Automatic Weather Station and Power Grid

GRIN Verlag

Bibliografische Information der Deutschen Nationalbibliothek:

Die Deutsche Bibliothek verzeichnet diese Publikation in der Deutschen Nationalbibliografie; detaillierte bibliografische Daten sind im Internet über http://dnb.d-nb.de/ abrufbar.

Imprint:

Druck und Bindung: Books on Demand GmbH, Norderstedt Germany
ISBN: 978-3-656-58447-6

This book at GRIN:

http://www.grin.com/en/e-book/266824/application-and-composition-observing-system-of-automatic-weather-station

APPLICATION AND COMPOSITION OBSERVING SYSTEM OF AUTOMATIC WEATHER STATION (AWS) AND POWER GRID (PGMIS)

Irfan Jamil*[1], Rehan Jamil*[2], Zhao Jinquan[1], Li Ming[2], Awais Ansar[3], Raheel Ahmed[3], Imran Jafar[3], Rafaqat Hussian[4], Rizwan Jamil[5]

[1]Department of Energy & Electrical Engineering, Hohai University, Nanjing, China
[2]School of Physics & Electronic Information, Yunnan Normal University, China
[3]College of Automation, Harbin Engineering University, Harbin China
[4]Department of Electrical Engineering, North China Electric Power University, Beijing
[5] Heavy Mechanical Complex (HMC-3) Taxila, Rawalpindi, Pakistan

ABSTRACT

This paper presents the compositions observing system and applications of AWS and PGMIS which are widely inaugurated in meteorological system respectively. The brief discussion is done on technical levels and control aspects of new technology for Automatic weather station and Power Grid Meteorological Information System. Controlled by electronics devices or computer, the automatic weather station automatically observes weather and collects and transmits data. AWS is usually composed of sensor, transmitter, data processing device, data transmitting device and power supply. The transmitter converts weather parameters sensed by sensor into electric signal then; data processing device will process these electrical signals and convert them into corresponding meteorological elements. Power Grid Meteorological Information System (PGMIS) refers to the meteorological information comprehensive platform of Power Grid applied in power grid corporations at all levels, which is also a professional application system in combination of meteorological information and production and operation of power grid. The system mainly provides timely and comprehensive meteorological information related to operation of power grid to realize monitoring, tracing, forecasting and warning of disastrous weather, and offers aid decision for load forecasting, economical dispatching and accident prediction of power grid.

KEYWORDS

AWS, PGMIS, Meteorological element, Sensor, Monitoring, Power Grid

1. INTRODUCTION

Automatic weather station (AWS) is a novel application of wireless sensor in the field of meteorological factor [1]. Based on the technical communication, the structure and operation of automatic weather station can be classified according to two different methods [2]. It is usually classified into real-time automatic weather station and non-real time automatic weather station in accordance with timeliness of data providing. Real-time automatic weather station: this kind of station can provide real-time weather observation data according to specific time. Non-real time automatic weather station: this kind of station only can regularly record and store observation data, but cannot provide real-time weather observation data [20]. The automatic weather station can also be classified into attended automatic station and unattended automatic station according to manual intervention condition of automatic weather station. The automatic weather station automatically acquires all or partial meteorological elements, such as air pressure, temperature, humidity, wind direction, wind velocity, rainfall, evaporation capacity, sunlight, radiation and ground temperature and so on, makes statistics of and encodes acquired weather data and transmits them to central station computer as per actual business demand or save them into local medium. Automatic weather station can be widely applied into meteorology, electric power, hydrology, agriculture, environmental protection, airport, warehouse, scenic spots, territorial resources and scientific research.

Grid monitoring and information service is a fundamental function of a grid system and Power Gird information security has been given particular attention [6], [9]. In Power Grid Meteorological Information System (PGMIS) which relates to the meteorological information comprehensive platform of Power Grid employed in power grid corporations at all levels and information source of power grid meteorological information system mainly comes from professional meteorological department, automatic weather station of electric power and lightning location system. The information includes historical meteorological data, real-time weather data, weather forecast product and meteorological disaster information. The storage, assimilation and sorting of multi-information is finished through development of data communication and background pre-processing module. As for communication safety protection, the power grid meteorological information system is located in safety area, which exchanges data with systems of other protection regions through network isolation device. In principle, the format of data file exchange adopts E language text.

2. Composition of Automatic Weather Observing System

The automatic weather observing system is automatic weather station in a narrow sense and automatic weather station network in a broad sense. The automatic weather station network is composed of one central station and several automatic weather stations through communication circuit, as shown in the right figure (form network through GPRS/GSM communication). The system implements data collection and processing through GPRS/GSM network [3].

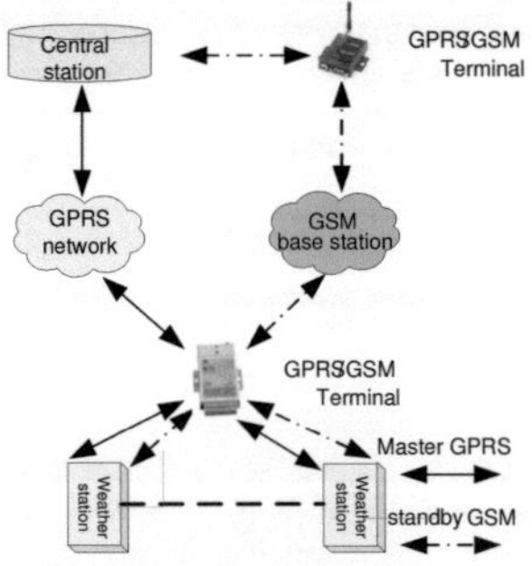

Figure 1. Network through GPRS/GSM communication

Automatic weather observing system mainly has following functions: automatically observe all meteorological elements, compile and store various kinds of weather report and observation data files, establish weather observation database, realize automatic transmission, calling and real-time control of weather observation reports and observation data files and remotely monitor operation status of system. From the structural diagram, each component of automatic weather station can be seen. The Fig.3 shows schematic diagram which provide a reliable means to calibrate AWS data-acquisition & procedure unit [4].

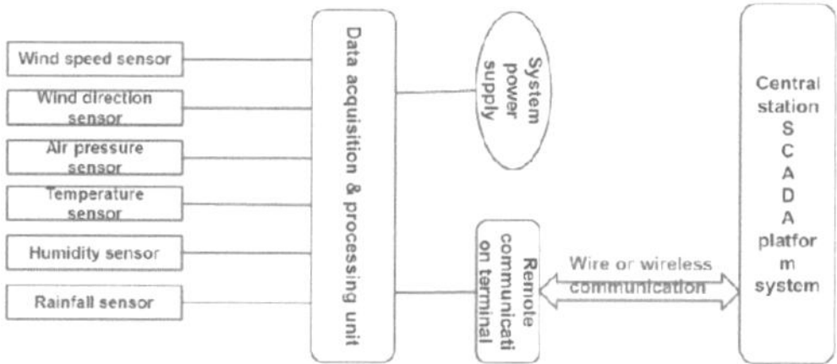

Figure 2. Schematic Diagram of Automatic Weather Station Composition

2.1. Meteorological sensor

Recently meteorological sensors have been actively introduced for climate monitoring [10]. Meteorological sensor can sensor the change of measured meteorological elements and convert them into useful output signal and is usually composed of sensitive element and converter [11]. The wind speed, wind direction, air pressure, temperature, humidity and rainfall sensors used for 6-factor automatic weather station are mainly explained here. These sensors can detect or forecast a wide range of phenomena such as fog, rainfall, floods, storms; aircraft wake turbulence and wind shear, as well as seismic and nuclear events [12].

- Wind speed & wind direction sensor: Wind speed and wind direction sensors vary depending on environment or business demand. These sensors identify automatic rain and wind measurement faults. At present, the mechanical wind speed and wind direction meter and ultrasonic wind speed and wind direction meter [5].

- Mechanical wind speed & wind direction meter: The wind direction part is composed of wind vane and converter. It is necessary to determine north firstly at the time of installation. When the wind vane rotates according to change of wind direction, Gray code or voltage signal is output in most cases. The sensing element of wind speed sensor is three-cup wind component. When wind cups rotate due to being affected by horizontal wind, frequency signals are output in most cases. The ultrasonic wind speed and wind direction meter transmits sonic pulse and measures the time or frequency (Doppler conversion) different of receiving terminal to calculate wind speed and wind direction. Digital quantity is output in most cases. The data interface includes RS232, RS485 and SDI-12. The ultrasonic type meter has no defect caused by friction loss of mechanical type meter.

- Air pressure sensor: Air pressure sensor converts change of atmospheric pressure into electrical signal and measures and processes electric signal through electronic measuring circuit to obtain pressure value. The commonly used electrical pressure sensor includes vibration cylinder pressure sensor and capsule capacitor pressure sensor. Data are usually output in form of analog quantity 4~20mA and digital quantity SDI-12, etc.

- Temperature sensor: The commonly used platinum resistance temperature sensor measures temperature according to the characteristic that platinum resistance varies with its temperature. When the measured medium has temperature gradient, the measured temperature is seemed as the average temperature of dielectric layer where temperature sensing element is. The analog or digital quantity for data acquisition and processing unit can be output through signal converter

- Humidity Sensor: The humidity sensitive element of humidity sensor mainly includes two types: resistance type and capacitor type. The humidity sensitive resistance refers to cover a film made of humidity sensing material on substrate. When water vapor in air is absorbed onto the film, both resistivity and resistance of element vary. When environment humidity change the dielectric constant of humidity sensitive capacitor also changes. And the capacitor variation is in direct proportion to relative humidity. The analog or digital quantity for data acquisition and processing unit can be output through signal converter.

- Radiation Filed: During construction of automatic weather station, the temperature and humidity weather sensors are usually installed outdoors. Therefore, they need to be protected with radiation shield (instrument shelter). At present, some sensor manufacturers have integrated temperature sensor and humidity sensor together and install them directly in radiation shield, as shown in figure below.

- Rain Fall Sensor: Various weather stations, hydrologic stations and environmental protection, agriculture and forestry defection departments use rainfall sensor to measure the rainfall of some place at some time and convert the rainfall into switching quantity, analog quantity or digital quantity signals which can be measured. The rainfall sensor includes single tipping-bucket type, dual tipping-bucket type and multi-tipping-bucket type, etc. In most weather stations, the rainfall sensor with 0.1mm resolution is used.

2.2. Data acquisition and processing unit

Data acquisition and processing unit is usually called by a joint name, its means data acquisition unit. As the fore of automatic weather station, it has the following main functions: data sampling, data processing, and data storage and data transmission. The automatic weather station is specially designed for the purpose of unattended station or for demand of weather monitoring in other severe conditions. Therefore, data acquisition unit shall have the following main characteristics.

- Micro-power since the automatic weather station is generally established in the field, no electricity is supplied to it in most cases. Therefore, the weather station must be of low power consumption

- Powerful remote communication capability with powerful remote communication capability, the automatic weather station can access common remote communication channels of domestic micro-power SCADA system, such as PSTN, GSM/SMS, GSM/GPRS, CDMA/1X, VHF, Inmarsat-C and Beidou and so on.

- Abundant sensor interface and high-precision acquisition provide abundant sensor interfaces, such as single/dual dry-reed relay interface (mainly used for tipping-bucket rainfall meter), 1-wire, SDI-12, RS485 and 4~20mA current loop and so on.

- High reliability: Most automatic weather stations are unattended stations. Therefore, its equipment must have high reliability. All elements adopt industry (or military) product and each sensor interface is configured with effective lightning protection measures to ensure timely and stable data processing of weather data observation specifications.

2.3. Power supply

The automatic weather station requires stable power supply system. Use mains supply If possible and carry out float charge for standby battery to supply power at the time of mains faulted. If computer is used, it is necessary to configure UPS and backup battery. In area without mains power, automatic weather station can be powered by battery. In this case, the battery can be charged with auxiliary power supply, such as diesel or gasoline generator, wind driven generator and solar panel and so on.

2.4. Communication terminal

The automatic weather station is mainly designed for unattended station. Therefore, its remote communication capability is very important. Problems as power, transmission cost and maintenance of terminating device have to be considered at the time of integration. Use VHF, GSM/GPRS, CDMA/1X, PSTN and satellite channel at the time of low transmission frequency due to large amount of data transmitted and adopt Ethernet, special line, optical fiber and other channels for data transmission at the time of high transmission frequency.

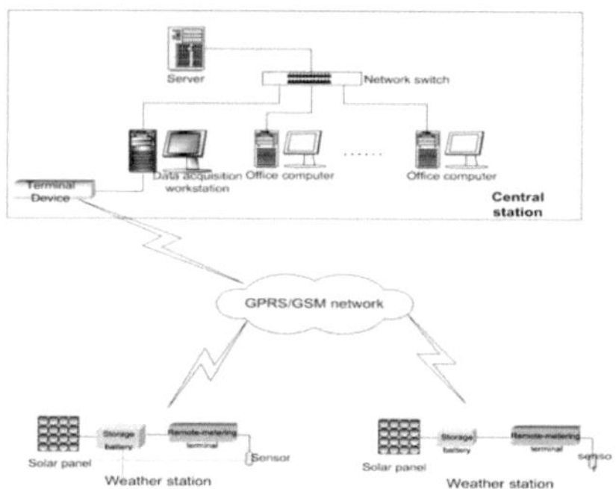

Figure 3. Structure diagram of GPRS Communication

2.5. Central station SCADA platform system

Supervisory control and data acquisition (SCADA) is mainly used for remote monitoring [13]. As business software, central station SCADA platform software is usually programmed according to demand of ground meteorological service. Its main functions include parameter setting, real-time data display, regular data storage and data maintenance, etc. The following functions

- Adopt modular design;
- Support multi-channel acquisition and multiple communication protocols of automatic weather station;
- Realize online configuration of operation parameters;
- View channel status and current data real-time information of original code and remote-metering station;
- Support data calling and time correction of remote-metering station. also can be added as per different service demands.

3. APPLICATION OF AUTOMATIC WEATHER STATION

Since automatic weather station is applied more and more widely, it mainly can be classified into following four types according to the construction purpose:

3.1. Standard automatic weather station

The standard automatic weather station is micro-power station designed according to ground weather automatic monitoring specifications with features of wide temperature range, low power consumption, high reliability and anti-interference, etc. This kind of weather station can measure air temperature, humidity, rainfall, wind direction, wind speed, air pressure, ground temperature, radiation, sunlight, evaporation and other parameters and also can access other meteorological sensors according to actual demand. The standard automatic weather station has to be established in a especially suitable place according to requirement of weather bureau, as shown in right figure.

Figure 4. standard automatic weather station

The standard automatic weather station processes data as follows:

The sampling rate of air temperature, humidity, air pressure, ground temperature and radiation is 6 times per minute. Remove the maximum and minimum values. Then, average the remaining 4 sampling values. The average value of 1 minute is instantaneous value.
The sampling rate of wind direction and wind speed is once per second. Determine the moving average values of 3 seconds, 2 minutes and 1o minutes. The average value of 3 seconds is instantaneous value.

The sampling rate of rainfall, evaporation and sunlight is once per minute.

3.2. Automatic weather station for anemometer tower of wind farm

The automatic weather station for anemometer tower of wind farm acquires wind speed, wind direction, temperature, humidity, air pressure and other data, providing real-time data source for wind power forecasting system. With this kind of station, users can conveniently know weather information of main windward side, facilitating output power adjustment of wind farm, which is significant for power balance and economical dispatching of power grid [19].

The automatic weather station for anemometer tower of wind farm processes data as follows:

- The sampling rate of air temperature, humidity and air pressure is 6 times per minute. Remove the maximum and minimum values. Then, average the remaining 4 sampling values. The average value of 1 minute before sampling is current instantaneous value

- The sampling rate of wind direction and wind speed is once per second and process data within sampling interval at the time of sampling. At present, the automatic weather station of anemometer tower carries out sampling and uploads data once every 5 minutes. The data of wind speed and wind direction includes instantaneous value, 5-minute average value, and standard deviation of wind speed, maximum value of wind speed and minimum value of wind speed. The average value of wind speed is arithmetic average value and that of wind direction is vector average value.

Figure 5. Automatic Weather Station for Anemometer Tower of Wind Farm

2.7. Automatic Weather Station for Status Monitoring of Power Transmission Line

The automatic weather station for status monitoring of power transmission line is applied to monitor the meteorological conditions of microclimate area of power transmission line with equipment installed on tower of transmission line, which can provide real-time weather data for disaster prevention and early warning of power transmission line [18]. This kind of weather station carries out real-time monitoring of wind speed, wind direction, temperature, humidity, air pressure, rainfall, optical radiation and other meteorological elements according to Q/GDW 243-2008 Technical Guideline for Weather Online Monitoring System of Overhead Transmission line and also can realize real-time video image monitoring of power line as required, providing reliable data base for status monitoring and maintenance of transmission line. Therefore, it is important means to ensure the safe operation of power transmission line. 00

The automatic weather station for status monitoring of power transmission line processes data as follows:

- The sampling and processing of air temperature, humidity, air pressure and radiation are same to those of standard automatic weather station.

- The sampling and processing of wind speed and wind direction are same to those of automatic weather station for anemometer tower of wind farm.

- The sampling and processing of rainfall and sunlight duration are same to those of standard automatic weather station.

2.8. Automatic Weather Station for PV Power Station

The automatic weather station for PV power station is applied to monitor the meteorological conditions in solar power plant, which samples total radiation, perpendicular incidence, scattering, wind speed, wind direction, temperature, humidity, air pressure, solar panel temperature and other parameters, providing real-time weather data for solar power forecasting system. With this kind of station, users can conveniently know weather information of PV power station, facilitating adjustment of output power, which is significant for power balance and economical dispatching of power grid[16].

The automatic weather station for PV power station processes data as follows:

- The sampling and processing of air temperature, humidity, air pressure and radiation are same to those of standard automatic weather station[17].
- The sampling and processing of wind speed and wind direction are same to those of automatic weather station for anemometer tower of wind farm.
- The sampling and processing of sunlight duration are same to those of standard automatic weather station.

Figure 6. The automatic weather station for PV power station made by NARI Group of Corporation, China

4. POWER GRID METEOROLOGICAL INFORMATION SYSTEM (PGMIS)

It is a comprehensive platform which combines meteorological information with productions and operation of Power Grid. It has been used in State Grid corporations for monitoring, tracing, forecasting and warning sever weathers. Also, it offers aid decisions for load forecasting and dispatching.

4.1. SYSTEM FUNCTION

In Data access and communication information source of power grid meteorological information system chiefly occurs from professional meteorological department, automatic weather station of electric power and lightning location system. The database of power grid meteorological information system adopts DB2 or domestic relational database (DM, Neusoft Open BASE, etc.). Its inner part can be logically divided into real-time database and historical database. The real-time database is used to save original meteorological data acquired from automatic weather station, including air temperature, rainfall, humidity, wind direction, wind velocity, air pressure and so on. The historical database is used to accumulate and save meteorological data, meteorological product (e.g.: cloud atlas and radar mosaic), meteorological processing data (e.g. static value of meteorological element), weather forecast, operation log and other text information. Besides, the historical database can save multiple forms of data, including numerical value, characters, text message, picture and binary files of other formats.

In Statistical analysis of data: Users can make statistics of meteorological element values of different areas in same period, sort gained results and save them into database; then, obtain the regional correlation of weather change in different regions through regional horizontal comparison and analysis, providing original data files for early warning of meteorological disaster. Compare and analyze accumulated value or actually measure value of meteorological elements in different periods (year, month or ten days) of same region to find out rules and difference of previous and later changes and analyze the immanent cause of weather phenomena and its possible influence on power grid. Calculate the current anomaly value of different weather in combination of historical average value, i.e. difference value between current element value and average value of many years. The year-on-year value and link relative ratio of weather elements can also be calculated. Analyze the weather features of current year, current month or current ten-day period; make statistics of occurrence probability and extreme value for historical data related to meteorological elements in each region and period and display them in form of graph and list, providing decision basis for medium and long term production and dispatching management of power grid.

4.2. GIS geographic information integration

The use of geographic information system (GIS) developed rapidly in the 1980s [14].Provide overlapping of vector layer and image data, including information of many important geographic positions, such as provincial boundaries, city regions, substations, power lines and automatic weather stations and so on, which are drawn respectively in form of point, line and plane, with additional function to give mark through text and simple geometric. The basic map operation can also be carried out on user graphic interface, including map view control (zoom in/out and translation and map layer display control (visibility with different scales and give mark or not. Users can define the drawing attributes of geographic elements through interface operation or editing configuration files, such as projection, title, scale, legend, line color, line type, area filling style and description text and so on with respect to geographic information systems [15].

4.3. PGMIS Application function

- Display of meteorological information: The Actual condition display of meteorological elements, Real-time monitoring of meteorological elements: display real-time metrological elements (including air temperature, air pressure, humidity, wind velocity, wind direction, rainfall and cloud cover) of important cities on map. Click the right key to display list query. Real-time query of meteorological elements: user interface displays data query table and graph (line graph or bar graph) and isothermal graphic display: display real-time isothermal on GIS graph and attach illustration for description.

- Statistical data display of meteorological elements: The statistical data query of meteorological elements: user interface displays data query table and graph (line graph or bar graph), graphic display of rainfall isoline: display historical rainfall isoline on GIS graph and attach illustration for description, Comparative display of real-time meteorological elements and historical average data: user interface displays the comparative analysis curve of real-time meteorological elements and historical average data.

- Display of forecast information: The table query of weather forecast data, the query content of weather value forecast includes the forecasted temperature, wind velocity, wind direction, air pressure, rainfall, humidity and total cloud cover of each hour in future 36 hours since user selects the time, the historical forecast data can be inquired, the query date can be set and display region refers to each prefecture-level city.

- Comparative analysis of temperature forecast graph: The monitoring of temperature value forecast graph: user interface displays multiple temperature graphs (line graph), including comparison of forecast temperature graph and actual temperature graph, display temperature forecasting value of each region in future 36 hours in form of graph, multiple region lists can be selected to realize comparison of regional temperature forecast graph and display region refers to each prefecture-level city.

- Display of forecast information: Display of weather forecast product, Display of weather forecast product includes text display and picture display. The displayed content includes conventional forecast, text forecast and early warning and city weather forecast, the static page displays the latest daily, weekly, monthly and quarterly forecast texts, Texts and text files of historical forecast can be inquired, Random time can be selected to inquire forecast product.

- Dynamic display of satellite cloud picture and radar image: Display the satellite cloud picture and radar image in past one day in the form of static picture. Display the satellite cloud picture and radar image in past one day through picture overlapping and auto play, Overlap satellite cloud picture and radar image on WEBGIS to realize regional location display. The satellite cloud picture and meteorological radar image of any time can be inquired.

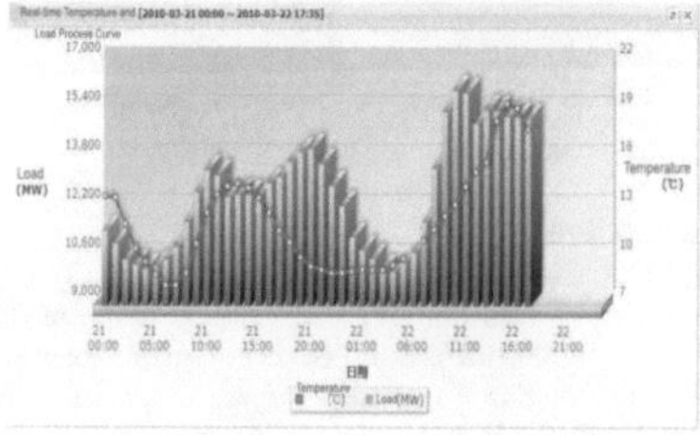

Figure 7. Comparative display of actual meteorological elements graph and power load

4.4. GIS Application in power grid meteorology

- Early warning of typhoon path and disaster influence: draw typhoon path and operation status of sometime precisely on WEBGIS, The process inversion of current typhoon and historical typhoon paths can be inquired. draw the forecast path of typhoon within 72 hours on each path point, draw actually influenced areas by force 7 wind and force 10 wind on each path point; and analyze the crucial power facilities in the area; realize early warning through list, warning mark and other forms, Path points of different colors stands for different wind forces, the travel process of typhoon path can be dynamically demonstrated and distance away from typhoon eye can be arbitrarily set on GIS graph and must be circled on GIS graph after setting.
- Early warning of extremely disastrous weather: Display the design data of power equipment (important substations and lines, etc.) related to resisting meteorological disaster on WEBGIS. When the corresponding actual meteorological data exceeds the limit value of equipment design data for disaster resisting, issue early warning in form of list and warning mark.
- Dynamically display various kinds of early warning information issued by meteorological department, display early warning which has been issued in scrolling way and GIS function expansion and Overlap the operation status of automatic weather station on WEBGIS, Cover the weather value forecast of 10KM * 10KM grid point on GIS graph Form layer of infrastructural items on GIS graph, including name, coordinates and other data and data at this layer can be maintained by user.

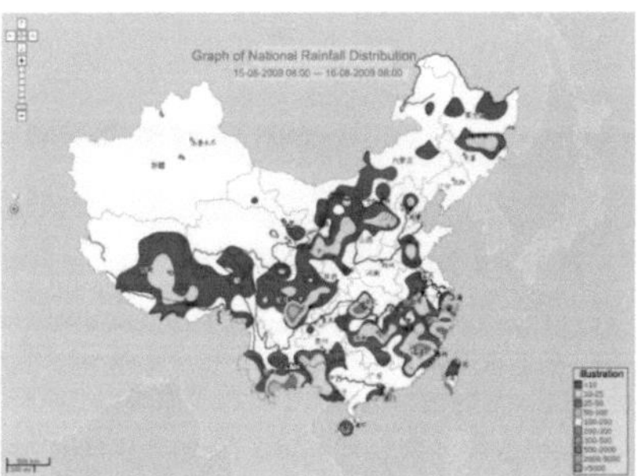

Figure 8. Rainfall Distribution of DWQX100 Power Grid Meteorological Information System (GIS) of NARI Group Corporation, China

4.5. References Application Effect of Power Grid Meteorological Information System

Power grid meteorological information system that includes program-level management skills is required [7]. It provides reliable basis of meteorological information for production, operation and dispatching of power grid, power marketing, power infrastructure construction and infrastructure management and offers powerful support for production, dispatching and decision of power grid. Power grid meteorological information system improve the warning and prevention capability of power grid to meteorological disasters and minimize the influence of meteorological factors on power grid, so as to ensure the effective, safe and stable operation of power grid. It reduces the power grid damages from meteorological disasters and improves the ability of disaster prevention and mitigation. Besides, the system also provides early warning of server weather such as typhoons, storms, rainstorm, lightning stroke, squall line wind, line pollution-flashover and line covered with ice.

3. CONCLUSIONS

The meteorological data are closely relative to our production and life and is extremely significant for all countries, enterprises or individuals. In the increasingly serious natural environment, meteorological data provides safety protection for us to prevent natural disaster and reasonably use natural energy. With scientific and technological progress, the construction and application of automatic weather station (AWS) and power grid meteorological information system (PGMIS) will be more advanced with more precise and timely data, making us have more progress in future communication and enjoy beautiful living environment. The brief discussion is done in this paper and also illustrated their principle, function application and also append throughout overview at technical levels of the system. The aim of this article is to guarantee the quality control of automatic weather station data & power grid meteorological information system [8].

ACKNOWLEDGEMENTS

The authors would like to acknowledgement technical support from the State Grid Electric Power Research Institute (SGEPRI), Nanjing Automation Research Institute (NARI) and financial support from the Hohai University, Nanjing, China.

REFERENCES

[1] Lee, S.hyun. & Kim Mi Na, (2008) "This is my paper", ABC Transactions on ECE, Vol. 10, No. 5, pp120-122.

[2] Gizem, Aksahya & Ayese, Ozcan (2009) Coomunications & Networks, Network Books, ABC Publishers.

[3] Ronghua Zhong, Shen Jun, Peng Xu "Analysis and de-noise of time series data from automatic weather station using chaos-based adaptive B-spine method" International Conference on Remote Sensing, Environment and Transportation Engineering (RSETE), PP. 4765 – 4769, June 24-26, 2011

[4] Jian-Ming Li, Han, Shu-Yong Zhen, Lun-Tao Yao "The assessment of automatic weather station operating quality based on fuzzy AHP" International Conference on Machine Learning and Cybernetics (ICMLC), PP. 1164-1168, July 11-14, 2010

[5] Xingang Guo, Yu Song "Design of automatic weather station based on GSM module" International Conference on Computer, Mechatronics, Control and Electronic Engineering (CMCE), Vol.5, PP. 80-82, August 24-26, 2010

[6] Wu Xiangjuan, Lv Wenhua, Xing Hongyan, Zou Yingquan "Research on temperature channel calibration of data-acquisition unit of automatic weather station" 10th International Conference on Electronic Measurement & Instruments (ICEMI), Vol. 2, PP. 194-197, August 16-19, 2011

[7] Hasu, V., Koivo, H. "Automatic Rain and Wind Measurement Fault Identification in Mesoscale Weather Station Networks" Instrumentation and Measurement Technology Conference Proceedings, IEEE, PP. 489-494, May 12-15, 2008

[8] Weibin Pei, Zhongliang Chen, Chunhao Feng, Zhi Wang "Design and Implementation of a Plain Grid Monitoring and Information Service" 5th IEEE International Symposium on Network Computing and Applications, PP. 277-284, July 24-16, 2006

[9] In-Ji Park, Run-Zhi Jin, Hyun-Ju Yang, Chang-Taek Hyun "A support tool for cost and schedule integration by connecting PMIS & PgMIS" International Conference on Engineering and Industries (ICEI), PP. 1-5, Nov. 29 2011-Dec. 1 2011

[10] Jing Li, China Jun Shen, An Chen "Multi-scale analysis for automatic weather station data" International Conference on Remote Sensing, Environment and Transportation Engineering (RSETE), PP. 4177-4180, June 24-26, 2011
[11] Wu Kehe, Zhang Tong, Li Wei "Research and Design of Security Defense Model in Power Grid Enterprise Information System" International Conference on Multimedia Technology (ICMT), PP. 1-4, Oct. 29-31, 2010
[12] Zhen Fang, Zhan Zhao, Xunxue Cui, Du, Lidong "Micro sensor network node design for meteorological parameter monitoring" IET International Conference on Wireless Sensor Network, IET-WSN, PP. 123-128, Nov. 15-17, 2010
[13] Mariño, P., Machado, F., Fontan, F.P., Otero, S. "Hybrid Distributed Instrumentation Network for Integrating Meteorological Sensors Applied to Modeling RF Propagation Impairments" IEEE Transactions on Instrumentation and Measurement, Vol. 57 Issue. 7, PP. 1410-1421, July 2008
[14] Paros, J., Yilmaz, M. "Broadband meteorological sensors co-located with GPS receivers for geophysical and atmospheric measurements" Position Location and Navigation Symposium, IEEE, PP. 134-141, 2002
[15] Ghosh, S.K "Changing role of SCADA in manufacturing plant" Industry Applications Conference on Thirty-First IAS Annual Meeting, IEEE, Vol. 3, P. 1565-1566, Oct. 6-10, 1996
[16] Basu, A. "Development of a marine geographic information system for various data analysis and data integration in the Hawaiian exclusive economic zone" Challenges of Our Changing Global Environment, IEEE Conference Proceedings, Vol. 1, PP. 146 – 153, Oct 19-12, 1995
[17] Buccella, A., Perez, L., Cechich, A. "GeoMergeP: Supporting an Ontological Approach to Geographic Information Integration" International Conference of the Chilean Computer Science Society, PP. 52-61, Nov. 10-14, 2008
[18] Ma Leiming, Pudong Meteorol. Bur, Geng Fuhai, Wuyun Qigige, Zhou Guanggiang "Numerical weather prediction in Yangtze River Delta region with assimilation of AWS and GPS/PWV data" Robotics and Applications (ISRA), 2012 IEEE Symposium on, PP.741-743, 3-5 June 2012.
[19]Wu, F Coll., Ju, P "Control strategy for AWS based wave energy conversion system" Power and Energy Society General Meeting, 2010 IEEE, PP. 1-2, 25-29 July, 2010
[20] Zhihui Huang , Xiaobo Wang, Shaodong Chen, Yijun Zhang "Analysis on the induced overvoltage generated by near triggered lightning in the AWS power distribution system" Electromagnetic Compatibility (APEMC), 2010 Asia-Pacific Symposium on, PP. 1522-1525, 12-16 April, 2010
[21] da Costa, Jose S, P., Sarmento, A., Gardner, F. "Modeling of an ocean waves power device AWS" Proceedings of 2003 IEEE Conference on Control Applications, Vol.1, PP. 618-603, 23-25 June, 2013
[22] Feng Wu, Xiao Ping Zhang , Ping Ju, Sterling, M.J.H. "Optimal Control for AWS-Based Wave Energy Conversion System" IEEE Transactions on Power Systems, Vo. 24, PP. 1747-1755, Nov. 2009

Corresponding Authors

Rehan Jamil* was born in Punjab province, City Multan, Pakistan on Feb 25, 1987. He received his bachelor in B.Sc. Electrical (Electronic) Engineering from Federal Urdu University of Arts, Science & Technology Islamabad Pakistan in 2009. Currently he is pursuing his Master degree at Yunnan Normal University, Kunming China. He is member of National Society of Professional Engineers (NSPE) and also member of American Society of Mechanical Engineer (ASME). He has been published 15 research papers and 2 e-books in referred international journals including IEEE and SCI conference during the period of master degree program from 2011- till present at Yunnan Normal University under the CSC Chinese Govt. Scholarship Funded program. He is involved in National and International Scientific and Technological Cooperation Projects of China (grant number: 2011DFA60460) at Yunnan Normal University. His research interest involves in Electronics, Renewable energy power generation.

Irfan Jamil* was born in Punjab province, City Multan, Pakistan on Feb 25, 1987. He received his bachelor degree in Electrical Engineering and its Automation from Harbin Engineering University, Harbin, China in 2011. Currently He is doing master degree and expected will be graduated in December 2013 at Hohai University, Nanjing, China. He did his master research as a Visiting Research Scholar in Tsinghua University, Beijing from 2012-2013 under the project No. is 2012AA050215 National High-Tech R&D Program (863 Program) of P.R China. He has been published 14 research papers and 3 e-books in referred international journals including IEEE conference during the period of master degree program from 2011-2013. He is registered member in Chinese Society for Electrical Engineers (CSEE), National Society of Professional Engineers (NSPE), International Association of Engineers (IAENG), Universal Association of Computer and Electronics Engineers (UACEE), Institute of Healthcare Engineering and Estate Management (IHEEM) and American Society of Mechanical Engineers (ASME). He is also editorial board of committee in International Journal of Innovation and Applied Studies (ISSR) and in World Academy of Science, Engineering and Technology (WASET). His research interest involves in Power electronics and Power system Automation.